Thank you for being our valued customer

We will be grateful if you shared this happy experience in the online review section.
This helpes us to continue providing great products
and helps potential buyers to make a confident decision.

Brand Name: Zebra Lines Publishing

NOTEBOOK

SIMPLE TO-DO CHECKLIST WITH 3 TOP PRIORITIES

THIS BOOK BELONGS TO

MONTHLY TO-DO LIST

JANUARY

FEBRUARY

MARCH

APRIL

MAY

JUNE

JULY

AUGUST

SEPTEMBER

OCTOBER

NOVEMBER

DECEMBER

Date

Top Priority

| 1 | 2 | 3 |

LIST IT OUT

Date

Top Priority

| 1 | 2 | 3 |

LIST IT OUT

- _____
- _____
- _____
- _____
- _____
- _____
- _____
- _____
- _____
- _____
- _____
- _____
- _____
- _____

Date

Top Priority

1	2	3

LIST IT OUT

Top Priority

| 1 | 2 | 3 |

LIST IT OUT

Made in the USA
Las Vegas, NV
10 March 2023

68837563R00070